THE

A, B, C

OF THE

BIOSPHERE

Professor Finch and his editor, Linnea Gentry, would like to thank
the many people who helped with this book, especially Lynn Ratener and Mary Beath.

Text Copyright © 1993 by Space Biospheres Ventures
Illustrations Copyright © 1993 by Mary Beath
All rights reserved.

No part of this publication may be reproduced or transmitted in
any form or by any means, electronic or mechanical,
including photocopy, recording, or any information storage
and retrieval system, without permission in writing from the publisher.

THE BIOSPHERE PRESS and its logo are trademarks of Space Biospheres Ventures.

Requests for permission to make copies of any part
of the work should be mailed to:
Permissions,
THE BIOSPHERE PRESS™
A Division of Space Biospheres Ventures
P. O. Box 689 • Oracle, AZ 85623 U.S.A.

ISBN 1-882428-11-0

Library of Congress Catalog-In-Publication data available upon request.

Printed in USA
by Arizona Lithographers

Printed on recycled and recyclable paper.

THE A, B, C OF THE BIOSPHERE

by Professor Finch

illustrated by Mary Beath

THE BIOSPHERE PRESS™

A is for Atmosphere, the air surrounding Earth

Above the Amazon and Africa and sunny Acapulco,
Among the antelope, the aspen, and the ambling armadillo,
Moves something quite important that I'd hate to be without,
Filled with lots of elements that life is all about.
I see it in the weather, as clouds up in the sky.
Without it fires cannot burn, the albatross can't fly.
So since you breathe it every day, it's not exactly fair,
When thinking of our atmosphere, to think of empty air.

B is for Biology, the study of life

"Now class, pay attention," said the bear to the birds.
"There is much for you to learn, so listen closely to my words.
Don't forget bacteria and their microscopic lives,
And mind the bees and the beetles and, of course, the butterflies.
Be courteous to the bats with whom you share the sky,
And never — no never — look a badger in the eye!
Thank the birch for its branches and wheat for grain and bran.
Beware the boa and the bobcat —
And especially, beware of Man!"

C is for Carbon, found in all living things

A coterie of camels and a choir of cats,
A company of cobras and a clutch of gnats,
A corps of creeping caterpillars, a crowd of caribou,
What is it that they all have and so do you?
In diamonds or in coalbeds, in clover or in clouds,
It's even in the silent corpse covered with its shroud.
Don't call it merely common, it's cosmic chemistry.
    In breathing and in eating,
        Carbon cycles through me!

D

is for Darwin, the great student of evolution

Did Darwin dream of dinosaurs
   as I like to do?
Some with frills, some with stripes —
   not with just a horn or two!
As he pondered their diversity
   and marveled at their size,
Did evolution's great detective
   suspect the means of their demise?
Or that the disappearance of dinosaurs
   by some catastrophic blow
Brought our timid mammal ancestors
   out of hiding long ago?
So many links to piece together
   in life's long mystery —
Even Darwin couldn't have known
   what might have been or is to be!

E is for Energy,
the mystery of power
in many forms

An elephant's trunk tears limbs off trees,
An eel makes shocking chemistry.
Windmill blades are spun by air,
And Einstein's E = M˙C squared.
Sunlight makes the eggplant grow,
Thermal force makes Mt. Etna blow.
Rushing water generates electricity —
All these are sources of energy!

F is for Food Web, the cycle that sustains us all

As the frog flicked its tongue and reached for the fly,
A ferret watched in shadow, hidden from view.
"Why bother with food," the ferret lamented,
"When I've heard that some day, I'll be eaten, too."
"Cheer up!" cried the vulture, his feathers a-flutter,
"Of this tried-and-true food web you shouldn't complain!
All fish, flesh, or fowl that's born must be nourished
So the cycles of life will always remain.
Be you fungus or fox, slow or fast on your feet,
    As each takes its turn,
      You become what I eat."

G is for Geology, the study of Earth's transformations

The science of geology is so deep you can mine it
With 4.6 billion years of history behind it.
Dig into limestone and delve into granite,
You'll read in the rocks the past of the planet.
Bones of ancient crocodiles lie in Greenland's ice
And remains of ancient magma metamorphose into gneiss.*
Huge roving continents cause strange transformations,
Such as India's gift to Asia of a new configuration —
    The shove that made the Himalayan wall
    And seeded seashells in Nepal.

* A metamorphic rock created from volcanic rock. Pronounced like 'nice'.

H is for Hydrogen, the most simple element

It's simple, it's light, it's the fuel of every star.
It helped create our habitat and make us what we are.
Oil, fat, and protein — this element they contain.
It's at home in hemoglobin, hormones, and hurricanes.

It can bond with other elements forming chemical duets,
Hook two of them to oxygen and H_2O* is what you'll get.
"You mean it's what I wallow in?" the hippo turns and asks.
"And it's helpful when I'm swallowin'!" the sly hyena laughs.

* water

I

is for Insects,
Earth's flying invertebrates

Call them weird, call them creepy,
   beyond the wildest imagination,
Insects are invertebrates
   worth our deepest admiration.
Compound eyes, three pairs of legs,
   and antennae in the air
Combine to give this form of life
   a very special flair.
Butterflies come from caterpillars
   and honey from the bee;
Silkworms spin us strands of silk,
   and ants recycle naturally.
Though some may think them irksome
   for their buzz, sting, or itch,
Imagine if they stopped their work,
   or their love of flowers quit!

J is for Jet Stream, a river of wind high in the sky*

Above a jaguar jogging through a jungle glade,
Above a jackrabbit crouching in a patch of shade,
Above a jellyfish drifting on the tide,
Above a jackdaw soaring o'er the countryside —
There's a river of wind about ten miles high
Whipping from out of the western sky,
Dividing the Tropics' balmy breezes
From the icy blasts of polar freezes.

* one in the Northern Hemisphere, one in the Southern

The five Kingdoms are: Plantae, Animalia, Monera (bacteria & blue-green algae), Fungi (molds & mushrooms), and Protoctista (seaweeds, watermolds, & several other water-dwellers). The classification system of arranging living things (called Taxonomy) was developed by Carolus Linnaeus in the 18th century. From the largest to the most specific, these taxonomic groups are: Kingdom, phylum, class, order, family, genus, species.

K

is for Kingdom,
the five main groups
of living things

"Come on," complained koala, "stop kidding me!
I don't get the hang of all this taxonomy.
If living things are separated into kingdoms five,
Then I'm kissing kin to one fifth of all that's 'live!"
"Correct!" cawed kingfisher. "Indeed, that's very true.
Although you seem quite different from kiwis or kinkajous.
Whether animal or mushroom, plant, bacteria, or mold,
Bred from egg, seed, or spore, be a species young or old —
"Stop! I'm so confused," cried koala with a sigh."
   "Be cool!" quipped kingfisher.
   "Just be glad we're classified!"

L is for Lithosphere, the bedrock below us all

The lizard lies upon the rock, with lichen 'neath his head.
Who would think that limestone could hold so fine a bed?
Under rocky peaks and chasms deep extends a base of stone,
The lithosphere, the solid crust, that shapes our earthly home.
The lobster lurks in watery brine, the loon takes to the air,
But lizards, leopards, you, and I require more solid fare.
Lucky for us that in eons past, lava cooled into crust
On which erosion worked away to make our pebbles, soil, and dust.

M is for Marsh, where land joins water

Out where land and water merge,
In lazy rivers and seatide surge,
Slow-moving water meanders and flows,
Making mud wherever it goes.
Mangroves in moonlight, cooters on logs,
Saltwater, freshwater, brackish bog,
The ocean's nursery, the waterfowl's nest,
The panther's refuge, where floodwaters rest.
A delight if you like a wet habitat,
Be you manatee, minnow, mouse, or muskrat.

N is for Nitrogen, an element used by all lifeforms

Do newts nibble nematodes at night along the Nile?
Is it normal for narwhals to wiggle when they smile?
And since nitrogen's invisible in our air all around,
How do microbes find it when it sinks underground?
Nasturtiums who need nitrates — or the nautilus and the tiger —
The nightingale who navigates from Norway to the Niger —
All this is natural history, and yet where would life be,
 Were it not for protein made
 From this gas you cannot see?

O is for Ocean,
Earth's vast body of water

"I have read in the sands,"
    said the octopus to the oyster,
"The secret of our origins
    in monera* long ago."
"There's much to be learned
    in the ancient ocean waters,
"And I," said the otter,
    "can confirm that it is so."
"Of course, you must be joking,"
    said the osprey overwrought.
"Monera as my ancestor?
    Most obviously not!"

* Monera = a cell or organism without a nucleus, such as bacteria or blue-green algae.
Pronounced mo-NAIR-ah.

P is for Photosynthesis, how plants harvest the sun

Far in the north with his close friend the puffin
A polar bear ponders the plan of the planet.
"If palms, pines, and poppies process the sunlight
And provide us with plentiful oxygen each day,
Preservation of plants seems a worthy endeavor
To prevent our own species from passing away."

"Indeed," sighed the puffin as she gazed o'er the ice,
"Though there's much on the Earth we'll never see,
Like Paris, the Pamirs, Peru, or Poughkeepsie —
It's a world worth preserving for plankton or people
 Where plants photosynthesize
 And polar bears philosophize."

Q is for Quest,
the search for knowledge,
full of adventure

A quest starts with questions
   when you simply must know,
Just how quick is quicksand?
   How big can squids grow?
Is the elegant quetzal
   on the verge of extinction?
What quirks in the Quaternary
   give it distinction?
From searching for quarks
   to exploring on Mars,
Our quest ends in answers
   that open the stars.

R is for Rainforest, lush tropical jungle

Where rain falls in torrents and great rivers run,
Where reptiles and rodents hide from the sun —
There's a rapture of species of marvelous shapes:
Tree-clinging orchids and the greatest of apes,
Ravenous raptors and night-roaming lemurs,
Amid rustles and trills, screeches and murmurs.
From the slow-moving sloth to the scarlet macaw,
The leafy green canopy shelters them all.
Enriching the Earth as part of life's plan,
   As we learn its rich secrets,
      We learn more for Man.

S
is for Savannah, a tropical grassland

Savannahs stretch out where warm suns glow,
With a few scattered trees, never touched by snow,
The tropical grasslands where humans first roamed,
Kin to the prairies in the temperate zones.
Savannahs can take fire, flood, and drought,
But when flames get hot, the snakes come out.
When raindrops drizzle and the grasses turn green,
Grazers and aardvarks fill the scene.
Though you won't see a starfish, a squid, or a shark,
Come visit any season — it's a natural park!

T is for Technology,
solving problems
by applying science

The tale of technology is a long one,
All these wonderful tools in our hands,
From the very first wheel to internal combustion,
And the first writing pens to computer programs.
I'm glad for tractors and vaccines,
The telescope, the loom, and the clock.
I'm fascinated by radar and huge submarines,
By the making of steel and the springs in a lock.
Transmitters, jet engines, and spaceships —
Where will they take us, what's it all worth?
Plastics, and rockets, and silicon chips —
Can we use them wisely without hurting Earth?

U is for Universe,
the totality of creation

V is for Vernadsky,*
the scientist
who discovered the
Biosphere Hypothesis

"Unless I'm utterly incorrect,
   the Universe is all that's known —
All galaxies voyaging through space,
   all meteors and moons."
Thus mused Vladimir Vernadsky,
   Russian science pioneer,
And turned his curiosity to Earth,
   home of the biosphere.

Unusual, perhaps unique,
   our planet is a world alive:
From vegetables to vertebrates,
   a multitude of creatures thrive.
What a wonderful volume of living things,
   the vicuña, violet, and vole!
Every part is a marvel
   and all the parts form a working whole!

* pronounced ver-NOD-ski

W

is for Wilderness,
what Man has left unchanged

Who needs the whale under watery waves?
Who needs the owl or the lion's roar?
What in wildness gives meaning to Man?
Who owns the wilderness, what is it for?

In the great North Woods of birch and pine,
Wolves hunt and wander in winters of white.
On Serengeti Plains in dry, dusty summers,
Wildebeest doze in a warm velvet night.

Out where the wind ruffles mane, fur, and feather,
Out in the ocean where seals ride the swells,
Listen for voices that call from the wild —
    What we leave to the wolves,
    We save for ourselves.

X is for X, the Unknown

"Yipes!" yelped the zebra at the zoo in Zaire,
"I've been here a year and I'm still not in gear.
I can't play a zither, do yoga, or yo-yo,
I don't know what X is, but I don't think it's zero."

Y is for Year, the cycle of seasons

"We're exceptional grazers, my boy," yawned the yak.
"That we can't work a gadget is not really a lack.
The world's so complex, it takes teamwork to balance.
   There's a need for us all,
      With our own special talents!"

Z is for Zoology, the study of animals

Well, that's a quick tour of the biosphere —
From algae to zebras, we're never alone.
So go out and enjoy what we hold so dear —
This extraordinary place we all call our home!

GLOSSARY

aardvark: an African mammal that eats ants and termites.

albatross: a large sea bird found mainly in the southern oceans.

algae: the simplest forms of green organisms, such as the green scum that forms on the sides of swimming pools.

bacteria: a large group of microscopic organisms with single-celled bodies, which live in soil, water, or in the bodies of plants and animals.

bog: a wet spongy area.

brine: water that contains dissolved salt.

compound eye: an eye made up of many simpler eyes that allow a wide field of vision, as in insects.

cooter: a turtle.

coterie: a social group.

Charles Darwin: a nineteenth-century British scientist who developed the Theory of Evolution in his famous book, *The Origin of Species by Means of Natural Selection*.

Albert Einstein: a German physicist who developed the Theory of Relativity explaining the relationship between mass and energy.

eon: one billion years

erosion: gradual wearing away of land or soil by the wind, rain, and other weather.

fungus: a simple organism, such as the mushroom, that lack the green-colored material of plants known as chlorophyll.

Galileo Galilei: an Italian astronomer who made the first astronomical telescope and studied the properties of gravity.

gneiss: a type of metamorphic rock that is similar to granite.

hemoglobin: the red substance in blood that carries oxygen from the lungs to the rest of the body.

hormones: chemicals produced in plants and animals that are important in growth and health.

internal combustion: the process inside an engine in which fuel is burned in a confined space to provide power.

invertebrate: any multi-celled animal lacking a backbone.

jackdaw: a black bird, similar to a crow, found in Europe.

kinkajou: a small flesh-eating mammal of Central and South America similar to a raccoon.

kiwi: a flightless bird the size of a chicken found in New Zealand.

lichen: an organism that is made up of green or blue-green algae and fungus living together in a mutually beneficial relationship.

limpet: a type of shellfish that clings to rocks or timbers.

loon: a fish-eating bird similar to a duck that lives in sub-Arctic and temperate areas.

metamorphose: to change in form geologically, not to be confused with metamorphosize which is how some animals, such as insects, change their form.

microbes: tiny organisms, such as bacteria, that can only be seen under a microscope.

monera: one of the five kingdoms of living things, comprised of bacteria and blue-green algae.

molecule: the smallest particle of an element.

nematode: a type of parasitic worm with a long unsegmented body.

newt: a small salamander that can live on land or in water.

Niger River: a major river in West Africa, 2585 miles long.

nitrate: the most common naturally occurring form of the element nitrogen

Pamir Mountains: a small mountain range along the north-east border of Afghanistan.

protein: a substance that the body needs to build muscle.

quark: a very tiny particle that is thought to be one of the building blocks of the atom.

Quaternary Period: the present geological period.

raptor: a group of flesh-eating birds such as a hawk or eagle.

Serengeti Plain: a large area of grasslands in East Africa.

silicon: a nonmetallic element that makes up over one fourth of the Earth's crust.

spores: the "seeds" that fungi and flowerless plants, such as ferns, grow from.

taxonomy: the system created by Carolus Linnaeus that scientists use to name organisms and organize them into groups.

temperate: the climate zone located between the tropics and the Arctic in both the northern and southern hemispheres.

transmitter: a device that sends radio signals.

tropics: the area flanking the equator usually associated with year-round warm weather and large amounts of rainfall.

Vladimir Vernadsky: the Russian geochemist who presented the idea that Earth is one huge, interrelated system.

vertebrates: any multi-celled animal with a backbone.

vicuña: a South American mountain mammal related to the llama.

vole: a burrowing rodent related to mice and rats.

wildebeest: an African antelope with a large ox-like head, also called a gnu.

Zaire: a nation in central Africa.

zither: a flat, stringed musical instrument.